WHAT IS MYCOLOGY?

Mycology is the study of **fungi,** a group of living things that includes mushrooms and yeasts. All mushrooms are fungi, but not all fungi are mushrooms!

The scientists who study fungi are called **MYCOLOGISTS.**

Words that are tricky to understand are in **bold.** Find out what they mean in the glossary.

Words that are difficult to say are in *italics.* Find out how to say them at the back of the book.

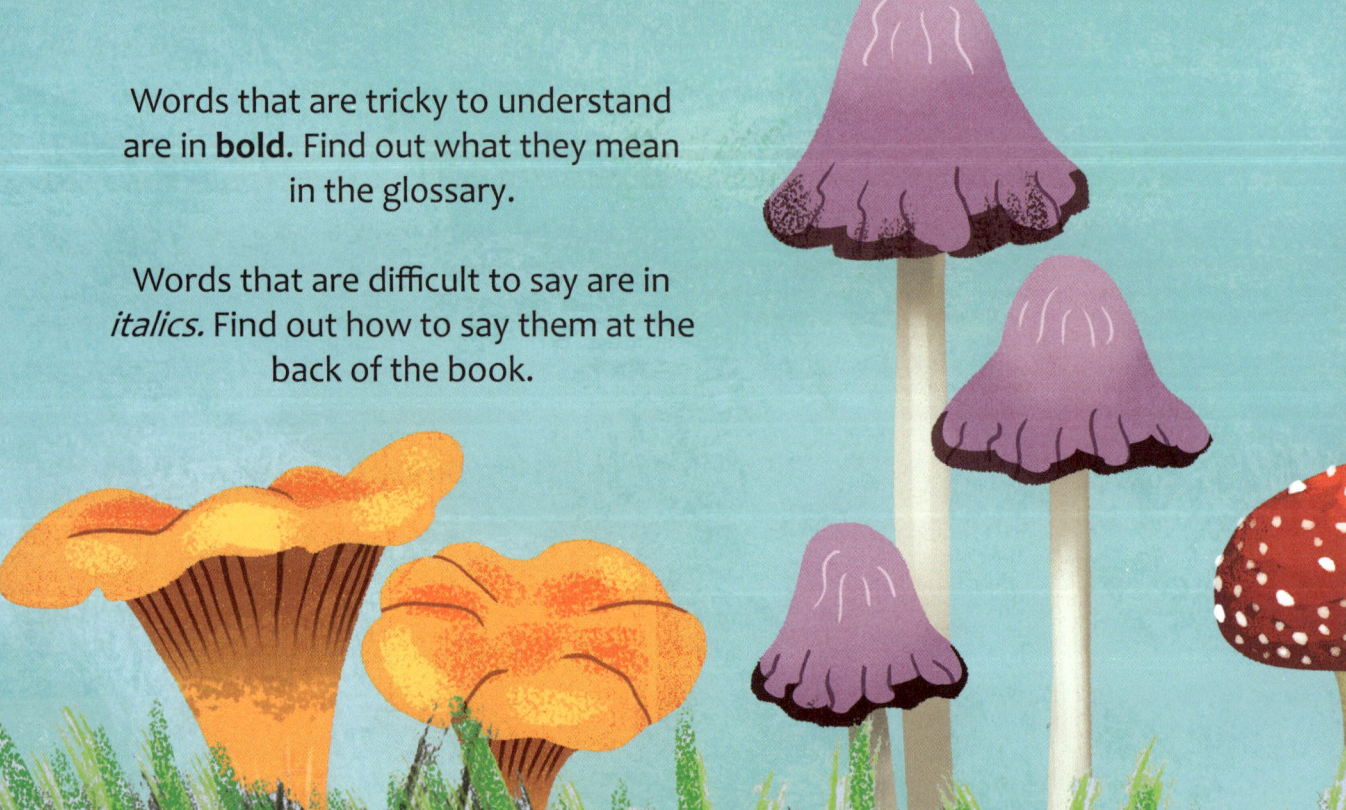

Can Mushrooms Save the World?

DISCOVER THE SCIENCE BEHIND MYCOLOGY
(my-KOH-loh-jee)

Written by Eliza Jeffery
Illustrated by Daniel Limon

More than 100 years ago, **fossil** hunters found strange stumps as big as tree trunks. Nobody knew what they were, until years later when **palaeontologists** worked out that they were actually **prehistoric** GIANT MUSHROOMS!

400 million years ago – before plants, trees, and land animals – monstrous mushrooms ruled the Earth!

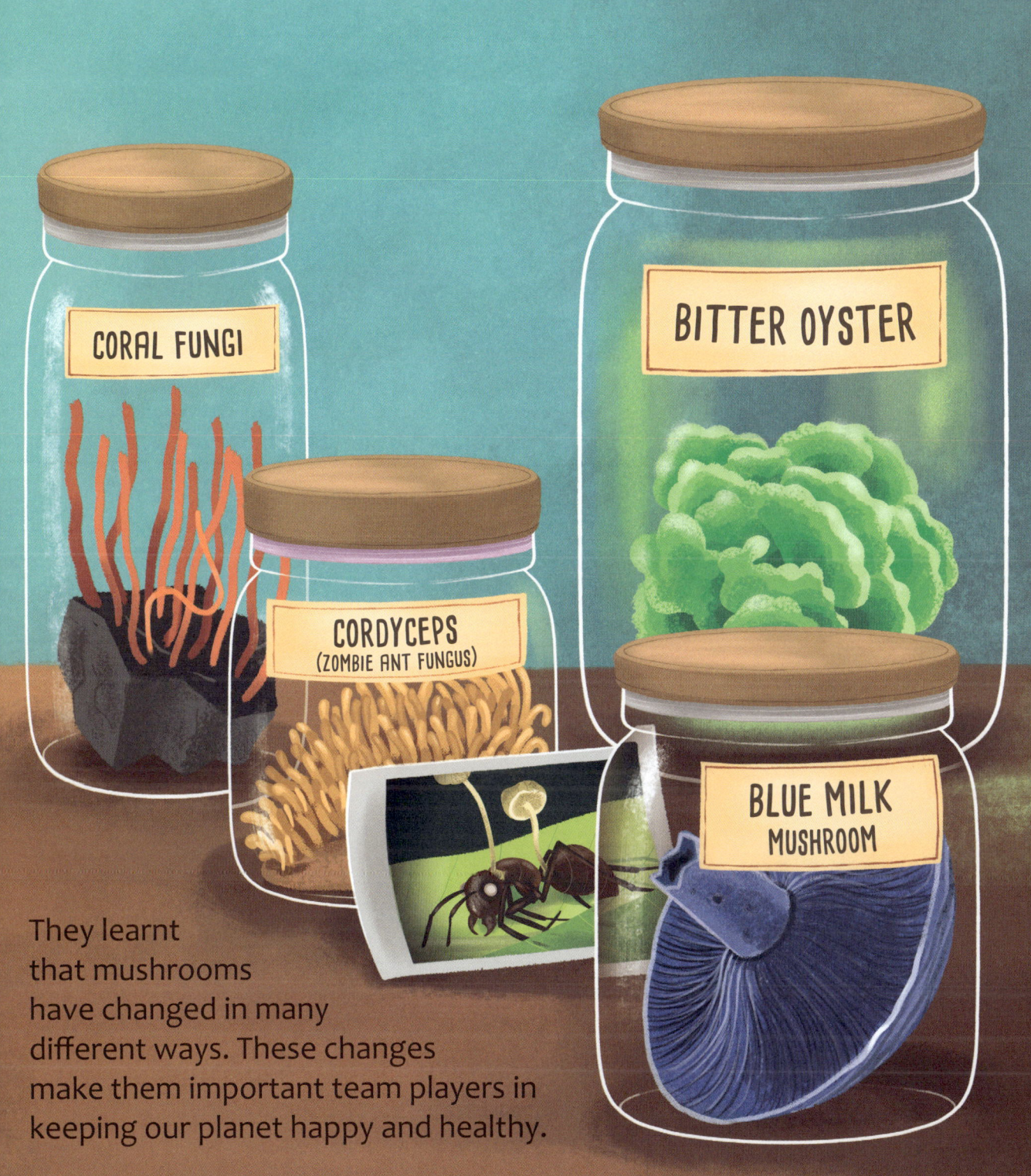

They learnt that mushrooms have changed in many different ways. These changes make them important team players in keeping our planet happy and healthy.

Fungi have a whole secret world hidden underground! They connect with trees through their roots. Trees and fungi make a great team, exchanging **nutrients** to keep each other healthy. Some scientists believe it's also like a natural internet network that allows trees to talk to each other.

But it takes a lot of mushrooms to help save the world! So, how do they spread? The answer is spores, tiny seeds that mushrooms release into the air. They are carried away by the wind or by sticking to insects and other animals.

Where spores land, new mushrooms grow! So mushrooms are popping up all over, even where there's no light...

Mycologists have discovered mushrooms that grow in complete darkness. Some of them even
GLOW IN THE DARK!

They look pretty cool. But that's not what it's all about! The glow is a genius way of attracting insects to come closer and help spread their spores.

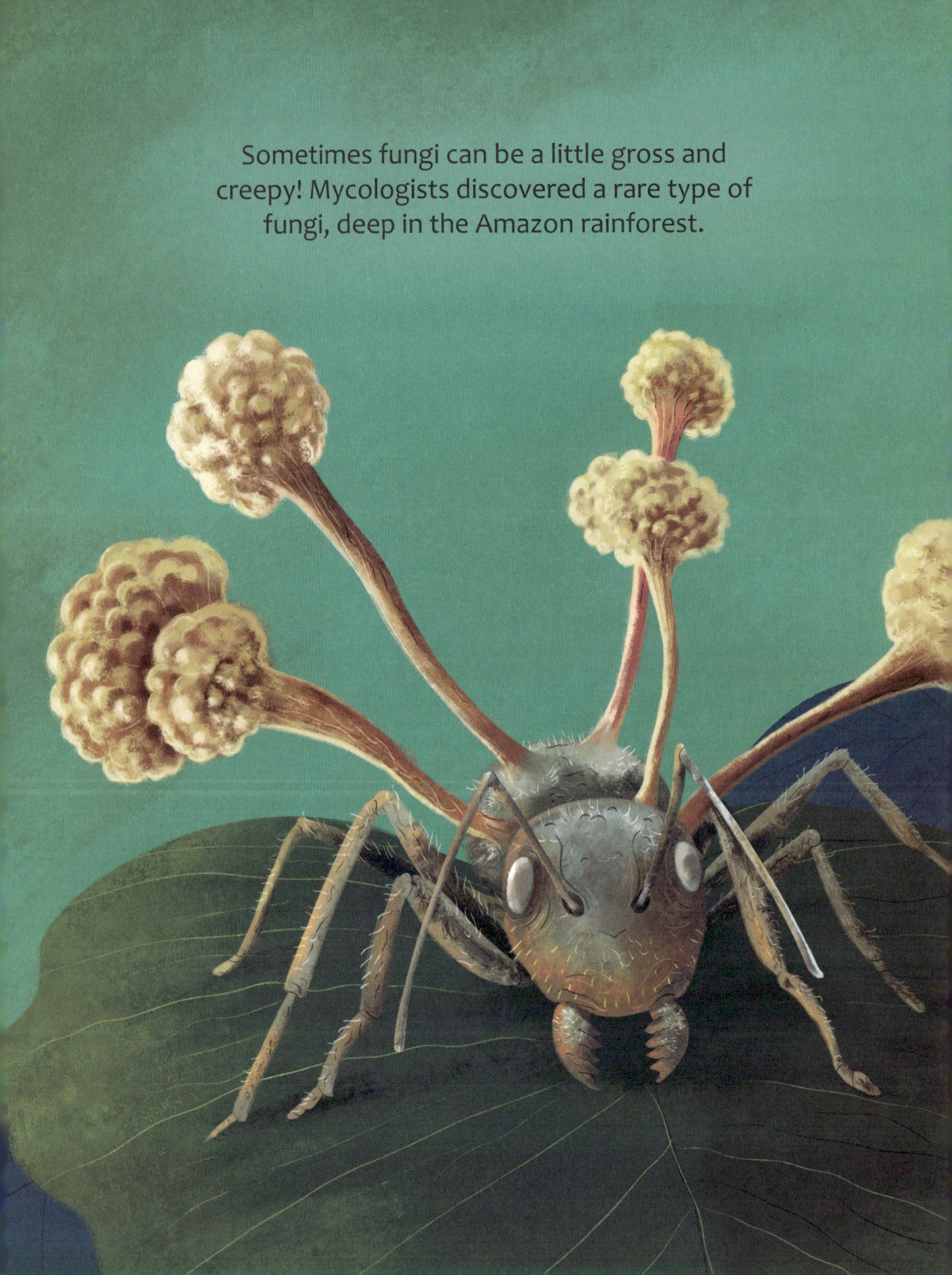

Sometimes fungi can be a little gross and creepy! Mycologists discovered a rare type of fungi, deep in the Amazon rainforest.

When the spores from these fungi land on forest ants, mushrooms take over their bodies, turning them into the living dead, real-life **ZOMBIES!**

Mushrooms may do some strange things, but they are more like us than you might think! They breathe **oxygen**, like us, and their **cells** contain a **substance** that's found in the **skeletons** of insects, crabs, and other animals.

Snails

Scorpions

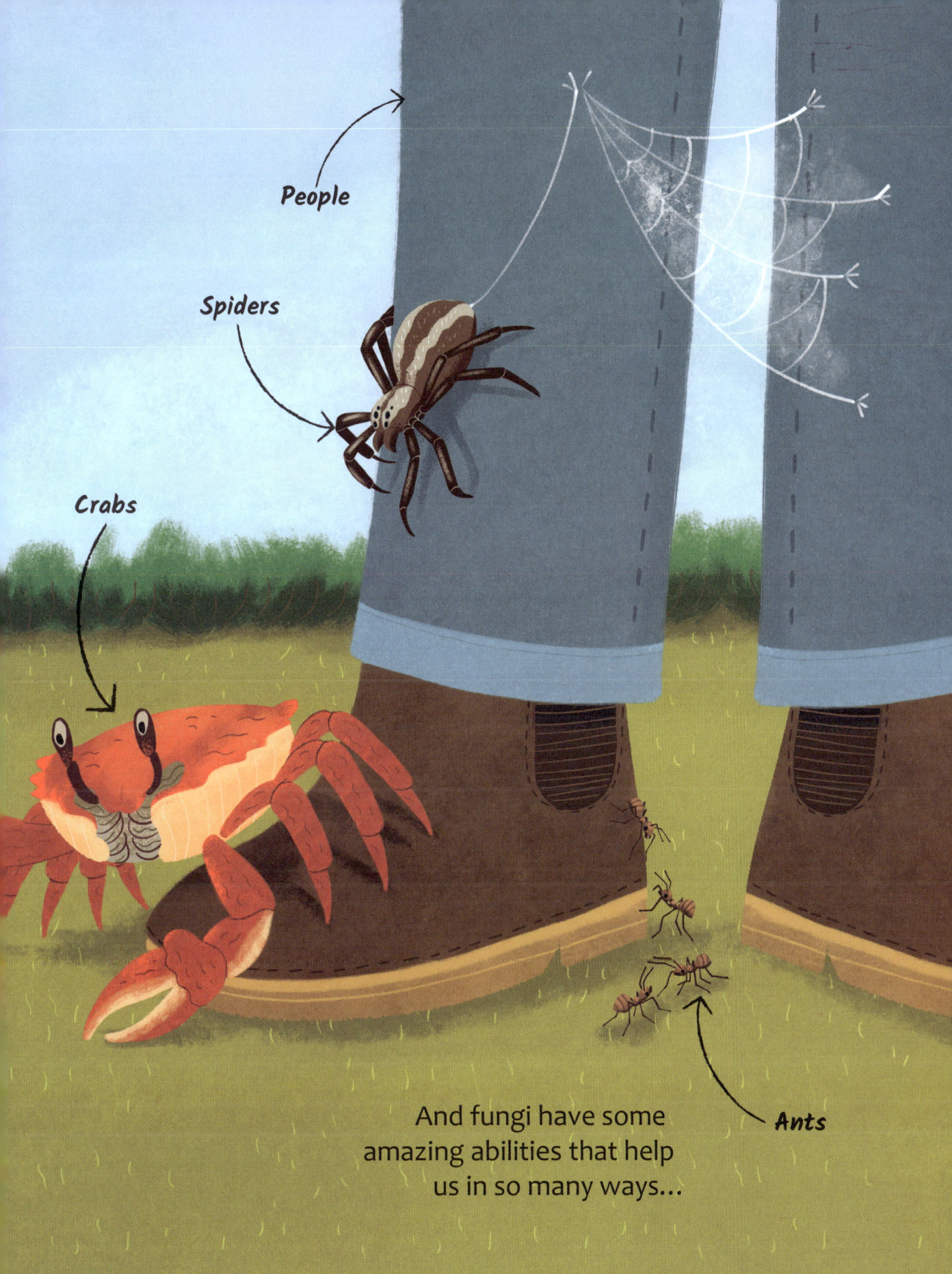

And fungi have some amazing abilities that help us in so many ways...

Mould is a special kind of fungi. It battles with **bacteria** for food, like **rotting** fruit. To win, mould makes a **chemical** that kills bacteria.

By studying this mould in a lab, scientists created a medicine called *penicillin* that treats infections caused by bacteria.

Mushrooms help the planet too; they are the perfect cleanup crew! Oil spills are bad news for wildlife. But mushrooms are like sponges that can eat up the harmful mess and break it down naturally.

Oil is used to make fuel, and fuel isn't very good for the **environment** either! So mycologists are using fungi that turn our waste into fuel for cars and trucks.

Using fungi for fuel could help us do less harm to the planet as we travel around. Who knows...soon we might all be driving around on mushroom power!

Oil is also used to make plastic. Plastic waste is a big problem, as it never really goes away. But fungi can be used to make everyday things like shoes and bags. Mushrooms are better than plastic because they **dissolve** naturally, keeping our planet clean!

Mushrooms can also help us just by being super tasty! They are great to eat on their own, but they can also be made to taste, feel, and even look like meat, just like a burger!

Using mushrooms instead of meat is an easy way to be kinder to the planet. This is because they need less water and land to grow.

Always remember to NEVER eat any mushrooms you find in the wild!

So, we now know that mushrooms can help save the world and mycologists are making that happen.

And they come in lots of fun and funky forms too…

Scarlet elf cup

Chanterelle

Black Périgord truffle

Shaggy ink cap

Fly agaric

Candlesnuff

Blusher

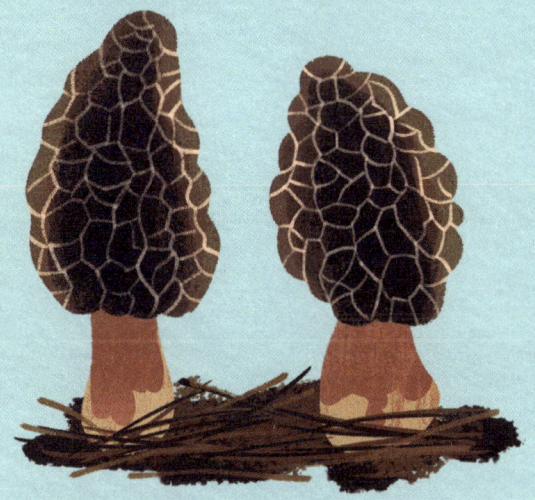
Morel

Yellow stagshorn

Amethyst deceiver

Mushroom
ANATOMY

With mushrooms saving the world, it's important for us to understand how they work. Let's explore the different parts of a mushroom and the functions of a fungi that make them fantastic!

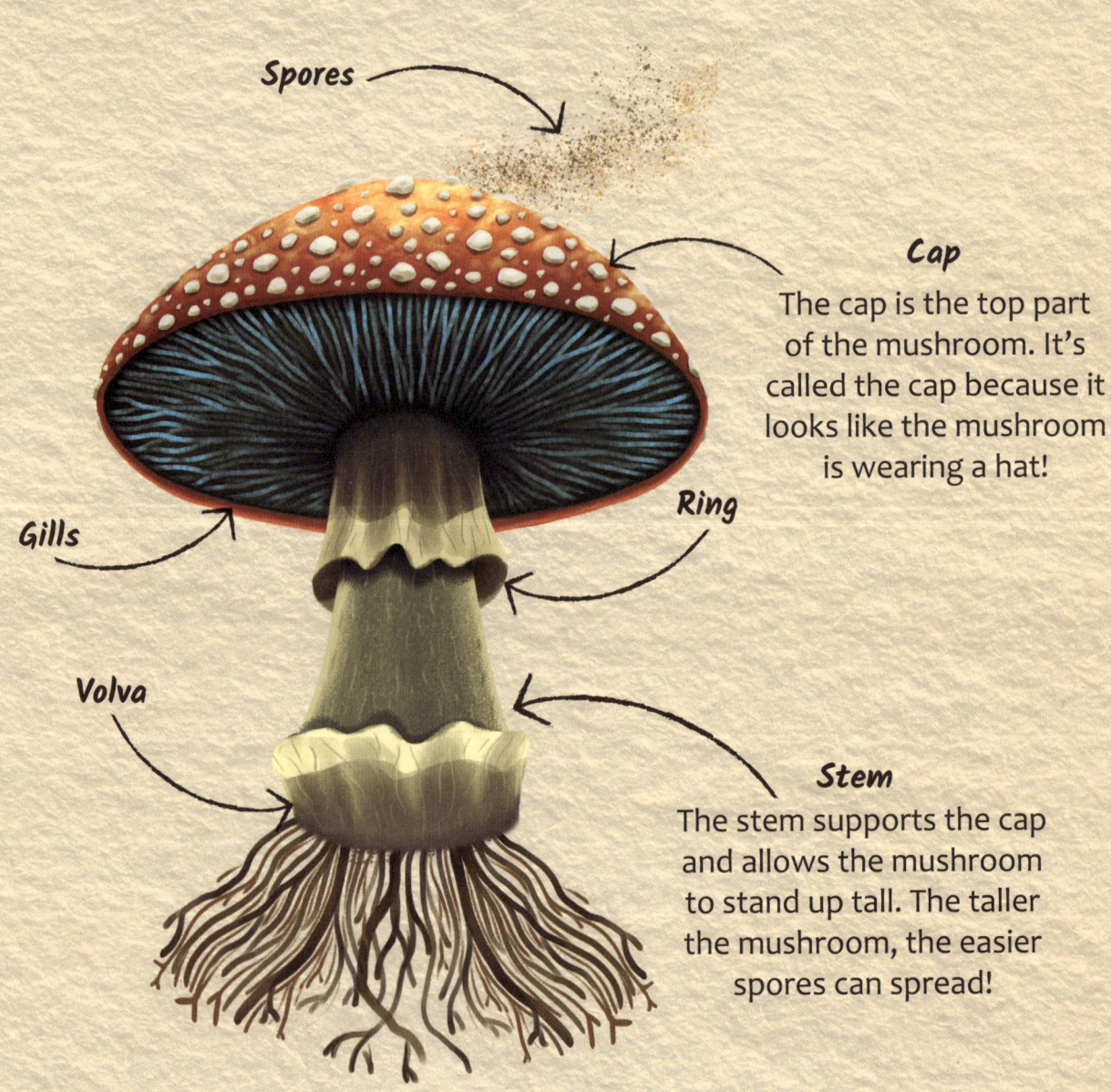

Spores

Cap
The cap is the top part of the mushroom. It's called the cap because it looks like the mushroom is wearing a hat!

Gills

Ring

Volva

Stem
The stem supports the cap and allows the mushroom to stand up tall. The taller the mushroom, the easier spores can spread!

Volva
The volva can be found at the bottom of a mushroom. It helps the mushroom to grow from the ground.

Gills
Gills are found underneath the cap of a mushroom and produce spores. Gills can look like teeth, needles or even like sponges!

Spores
Spores are tiny, dust-like seeds that shoot out and spread to become new mushrooms. The more spores, the more mushrooms!

Ring
The ring of a mushroom protects the gills so they can make spores. Mushrooms have **adapted** over time to keep themselves safe in this way.

Fun FUNGI FACTS

There's so much to discover about the world of mycology. Did you know some of these amazing fungi facts?

OUR FEET CONTAIN OVER 200 TYPES OF FUNGI

It can grow between our toes, on the heels of our feet and even under our toenails!

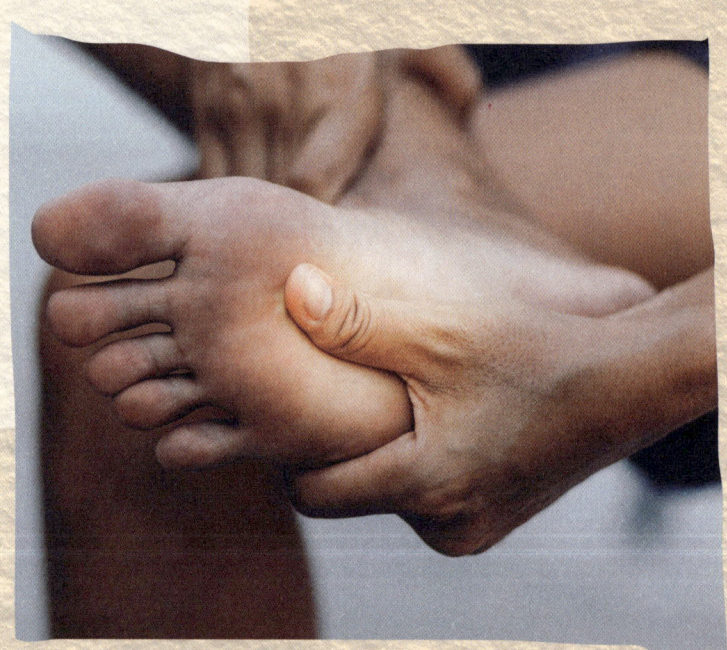

FUNGI ARE THE LARGEST LIFE FORMS ON EARTH!

The biggest one is nicknamed 'Humongous Fungus'. It doesn't look big, but its roots are all connected together, making it a huge fungus!

THE SHAGGY INK CAP MUSHROOMS EAT THEMSELVES!

These mushrooms melt into a black inky mess 24 hours after they've been picked!

MUSHROOMS HELP IT RAIN IN TROPICAL RAINFORESTS!

When mushrooms release spores, they attract **moisture** which then turns into raindrops. 90% of rain in tropical rainforests contain mushroom spores!

LIGHTNING GIVES MUSHROOMS THE POWER TO GROW!

Scientists have discovered that when lightning strikes mushroom crops, the speed of the growth more than doubles!

GLOSSARY

Adapted – when a living thing has developed special features or skills to help it survive in its environment (see below).

Bacteria – tiny living things that can be found in all natural environments (see below).

Cells – the smallest parts of a living thing.

Chemical – a substance made up of the same things, often made by humans.

Dissolve – when a solid mixes with a liquid, like water.

Environment – everything that is around us.

Fossil – the remains of plants and animals that lived long ago.

Fungi (the plural of *fungus*) – living things that are neither plants nor animals.

Moisture – small amounts of water or other liquid present in air or soil.

Nutrients – substances or ingredients that plants and animals need to live and grow.

Oxygen – an invisible gas in the air that plants produce, and people and animals need to breathe.

Palaeontologists – scientists who study fossils (see left).

Prehistoric – the time before humans existed.

Rotting – breaking down and decaying.

Skeleton – the bony frame that supports and protects the body of a person or animal.

Substance – the stuff or material from which something is made.

Zombies – fictional creatures that have risen from the dead.

HOW DO I SAY?

Anatomy
uh-NAT-uh-mee

Mycologists
my-KOH-loh-jists

Mycology
my-KOH-loh-jee

Penicillin
pen-uh-SIL-in

THE BIG QUESTIONS ANSWERED

This is more than just a series of books; it is a complete resource.
Accompanying each book is a variety of FREE material to engage curious kids with science.

 www.thebigquestionsanswered.com

Use the QR code to visit the website, download free resources, and discover other books in the series.

On the website, find out incredible things about mycologists, including what they do, some of their greatest discoveries, and what it takes to become an expert in this field of science.

The material is also available for home or classroom use, supporting all the information in this book.

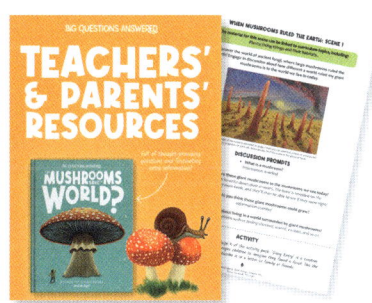

Teachers' & Parents' Resources
With discussion prompts and questions, extra information, and facts around key topics.

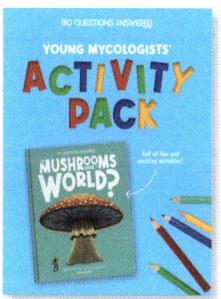

Young Mycologists' Activity Pack
Fun activities for wannabe mushroom experts, including creative writing, drawing, word searches, and much, much more.

Beetle Books is an imprint of Hungry Tomato Ltd.

First published in 2024 by Hungry Tomato Ltd
F15, Old Bakery Studios, Blewetts Wharf, Malpas Road,
Truro, Cornwall, TR1 1QH, UK.

ISBN 9781835691267

Copyright © 2024 Hungry Tomato Ltd

No part of this publication may be reproduced, stored in a retrieval system, or transmitted in any form or by any means, electronic, mechanical, photocopying, recording, or otherwise, without prior written permission of the copyright owner.

A CIP catalog record for this book is available from the British Library.

With thanks to:
Editor: Millie Burdett
Editor: Holly Thornton
Senior Designer: Amy Harvey
Tim Cook for his valued contribution
The team at Beehive Illustration

Printed and bound in China.

Picture Credits:
(t = top, b = bottom, m = middle, l = left, r = right)
Shutterstock: Alexey Stiop 35bl; Anut21ng Stock 34mr; Badnews86dups 34bl;
Dalapix 33br; Edwin Butter 35tl; Godi photo 33bl; Kichigan 35mr; Lazy Panda 33tl;
Tintila Corina 33tr.